Carpal Tunnel Syndrome 90% Misdiagnosed:

For the Patient & Provider

Based on Clinical Research

Second Edition

Angela Rahn, MPT

Roger S. Rahn, MT, DC

Roger S. Rahn Chiropractic Professional Corporation

777 Minnewawa Ave., Suite 9, Clovis, CA 93612

Phone: (559) 324-0628

Fax: (559) 324-0506

Web site: www.DrRogerSRahn.com,

E-mail: roger@DrRogerSRahn.com

Carpal Tunnel Syndrome is Secondary to Undiagnosed Shoulder Problems 'TOS' 85 Patients Treated

A Reproducible, Inexpensive, Effective Non-Surgical Treatment For Peripheral Nerve Compression 100 Patients Treated

May this book be dedicated to two truly great & selfless humanitarian teachers:

Shri Sathya Sai Baba

and

Swami Turiyasangitananda.

Any good qualities I have are due to your patient teaching and perfect examples. I humbly offer my thankfulness.

Acknowledgements:

This second edition was completely re-formatted and edited by a very competent medical editor, Crystal Phend. We highly recommend her services and are very grateful for her cheerful participation. She can be contacted at C_Phend@nasw.org.

Additional editing provided by my friend,

Kent Hamlin of Fresno, CA.

We are grateful for your clear and precise command of written communication.

Testimonials

“I nearly had to quit my work as a massage therapist. My hands hurt and were numb all the time. In a few treatments my problem completely resolved.” --Donna Rhodes

***

"I had to leave work last Saturday with severe hand pain and numbness. With only 2 treatments my hand pain is gone and I have returned to work!" --Trudy, waitress

***

"I am a massage therapist at my own business. I was afraid that I would have to stop working due to numbness, pain and weakness in my hands. I no longer enjoyed my work – I couldn’t help people. I was skeptical at first but within 2 treatments I felt so much better – the pain was gone." --Sandy, massage therapist

***

"I have worked as a massage therapist for years and I love to help people. I had foot and leg pain. Also, my hands were very painful. I send my toughest clients to Dr. Rahn to get what I have – results. Neither chiropractic or massage helped – I tried it all. Actually, almost everyone where I work is going to take a series of seminars with Dr. Roger to learn what he does – I already tried some of what he does on chronic patients and they are getting better." --Lois, massage therapist

***

"I have been on disability twice, for years, for low back pain and also neck/shoulder/arm pain. The MRI was negative. Dr. R's treatments were not just good, they were Fabulous! I felt a little difference after one treatment; but, after 2-3 treatments I was feeling like I could go back to full time work in the career I had given up!" --Donna, paralegal

***

"Just want everyone to know that when I first visited Dr. Rahn, it was for a back ache...and in the process, I explained to him that I had a lot of pain in my knees. Several years ago, I fell and suffered a torn meniscus, which was removed through arthroscopic surgery. Unfortunately, over time, there have been further problems resulting in a situation that has led my orthopedic surgeon to recommend total knee replacement of both knees. When I first visited Dr. Rahn, I was taking Celebrex for the pain, limiting my time on my feet and sitting...as both were painful and tiring. That was about 5 months ago. Today, thanks to regular deep tissue massage and gentle chiropractic treatments...coupled with meditation...I am no longer taking Celebrex and my mobility is increased in length of time and type of activity. I continue to postpone knee replacement surgery...as I am hoping to continue to enjoy my present state of comfort and strength for as long as possible. At any rate, I am firmly convinced that Dr. Roger Rahn's holistic approach is the way to go in treating pain and injury. Dr. Rahn has developed an approach to treatment that considers all aspects of the human system....and I highly recommend consulting him in the course of treating carpal tunnel syndrome, backache and knee problems. Personal experience has taught me that his approach to treatment not only makes sense...but works!"

--Anonymous Web site guest, March 4, 2004

***

Contents **Page**

What is Carpal Tunnel Syndrome 12

Dr. Rahn's Perspective 23

Dr. Rahn's Thoughts on Chiropractic 26

Why Most Carpal Tunnel Syndrome Diagnoses Are Inaccurate 30
(this section is written primarily for health care providers)

Making a Correct Diagnosis 36
(with symptom survey)

Treatment Protocol for Non-Surgical Neurovascular Decompression 42

Satisfaction Guarantee 57

Research Supporting Our Treatment Protocol *(this section is written primarily for health care providers)* 59

Appendices

A: Ergonomic Considerations
B: Stress Reduction Considerations
C: Dr. Rahn on Nutrition
D: Technique for Neck Pain and Headache
E: Technique Applied to Leg Pain
F: Dr. Rahn's Hand Exam Form
G: Equipment

H: Carpal Tunnel Height Measuring
Procedure
I: Works Cited
J: Stretches and Exercises for TOS and CTS
K: About Dr. Rahn

What is Carpal Tunnel Syndrome?

Many people suffer daily with what they have been told is carpal tunnel syndrome. As a result of their pain and disability, some people have to retrain and work in a different capacity or retire altogether. At the very least, they have to work with a modified or light-duty work assignment. Some individuals with carpal tunnel syndrome have tolerated their symptoms for so long that taking pills when they wake is automatic. There are even those people who believe their problem has subsided because the pain has stopped, which is an incorrect conclusion resulting from the brain adapting to a chronic problem.

Carpal tunnel syndrome (CTS) is a problem with the nerves, muscles, and joints. This category of problems is usually referred to medically as musculoskeletal or neuromusculoskeletal problems. CTS is currently **one of the most common health care problems today**.

From an economic point of view, this problem is huge. For example, carpal tunnel syndrome (CTS) is one of the most common problems, and the hardest to diagnose and treat. Some reports tell us that it costs $30,000 to treat a case of CTS. In 1993, the U.S. Department of Labor reported the annual cost to treat CTS is about $5,000 per patient. Some reports currently tell us that treating a case of CTS can average between $6,000 and $10,000, depending on whether one or both arms are involved. It's a bold statement, but we believe that if these problems were treated as described in this book, it would be possible to balance California's budget.

Our treatment protocol is highly successful in eliminating the cause of the pain and disability of CTS. We received an E-mail from a woman in another state. In a plea for help, she shared that her hand pain was so great she could not sleep at night. She also stated that she was disabled and could not pick her children up. With a copy of our treatment protocol, her local massage therapist treated the front of her shoulders and completely resolved her problem.

Of course, every patient should be examined by a qualified health care provider, because only a provider can properly evaluate and rule out other threatening causes. Also, we feel a physical therapist or chiropractor, certified in this technique and listed on our Web site, is the best-qualified person to evaluate and manage a potential case of CTS. Lastly, a physical therapist, chiropractor or massage therapist should actually perform the treatment.

How wonderful that a disabled person with constant acute pain is now living pain-free. And, we never saw this patient! If our treatment is so simple that a massage therapist can learn it from an E-mail, why are other health-care providers not using our technique? There are two parts to this answer.

First, doctors diagnose carpal tunnel syndrome as a problem originating in the hand. The pain, weakness and tingling all are located in the hand. There is swelling in the hand. Often, there is compression of the nerve in the wrist (the median nerve at the flexor retinaculum). However, our research has shown that carpal tunnel syndrome is usually caused by muscle spasm at the front of the shoulder.

A similar example is when a pregnant woman has swelling in her ankles. The weight of the fetus on the pelvis compresses blood vessels causing decreased circulation and swelling which is then pulled down to the ankles by gravity. Likewise, tightness on the front of the shoulder puts pressure on blood vessels, thereby decreasing circulation. Since the pressure decreases the efficiency of the circulatory system in the arm, gravity keeps excess fluid in the wrist, which we see as swelling. It is the swelling in the hand that creates a bottleneck and compression of the median nerve. This spasm also puts pressure on the nerves at the front of the shoulder. Medical tests such as X-ray cannot identify muscle spasm in the front of the shoulder.

Anti-inflammatory medication may help. Painkillers will decrease the pain allowing the patient to work and function more normally. A surgical release that cuts the ligament that we have been talking about decreases pressure on the nerve and helps the overall problem. However, if part or all of the original problem originated at the shoulder, all of these treatment techniques will either fail or have poor results.

The second reason why health care providers do not utilize our technique is that our treatment requires considerable time and labor on the part of the health care provider. Our technique is a money-saving technique, not a money-making technique. With it, massage therapists or physical therapists are required to provide 30 minutes of hands-on care. It's easy to understand that health care providers can see more patients by simply dispensing medicines rather than by dedicating 30 minutes to hands-on labor.

Let's discuss conventional care. We already mentioned that anti-inflammatory medicine and painkillers might help with the symptoms. However, it's easy to see that this conservative treatment is dealing with the symptoms and not the cause of the problem. Anti-inflammatory steroids, either oral or injected, are often used to decrease pain and swelling. However, steroids have been shown to be ineffective as a long-term treatment for this condition.

The surgical release research has reported high rates of success with surgery. One study reports that about one third of patients treated with surgery complain of pain at the point of incision, and it takes about 9 to 10 months to reach the point of greatest benefit (the least amount of pain) following surgery.

Furthermore, surgery has limited success. Two years after surgery, about half of the patients report a return of original symptoms (pain, numbness and tingling). Many of these patients report weakness and consequently many must work with a modified workload or undergo vocational rehabilitation, which is especially true for patients doing heavy lifting and working with power tools. Additionally, the strength of the wrist is compromised. Altogether, we do not feel that surgery is a good alternative for most patients.

Often, anti-inflammatory medication is used first in an attempt to cure the problem. However, steroids only mask the symptoms. By the time this short-term treatment fails, more permanent damage may have been done. If the surgery is postponed more than three years after the original diagnosis, there is such a large amount of nerve damage that surgery is half as likely to be successful.

In summary, the medication and steroids are effective for short-term care but do not cure the cause of the problem. Surgery has limited success. We are suggesting that the main reason none of these treatments are successful in totally curing the presentation is because the problem usually originates at the shoulder.

The Causes of Carpal Tunnel Syndrome:

Past generations did not suffer as we suffer today. Why is this? Probably the largest factor is that workers of the past performed a broad range of work activities that involved a diversity of work postures and wide range of movements. Today's 'static' work posture and highly specialized, repetitive work environment creates the high occurrence! Although many researchers have postulated that CTS is primarily a result of repetitive movements, this is only part of the problem.

If the shoulders are elevated by a fraction of an inch, muscles are supporting the weight of the shoulder and arm, rather than the joint and ligaments. If the shoulder is not elevated, even if the shoulders are not moving, **tension in the muscles force the muscles to do (physiological) work**. This invisible work fatigues the muscles, creates muscle spasm, decreases circulation and may eventually shorten the muscle with fibrous tissue. Holding a piece of paper in front of you doesn't sound like much work, but don't forget about the weight of the arm. Think of your arm like a wrench or lever. Loosening a tight bolt requires less work with a longer wrench. Similarly, there's a huge difference in the amount of work a person does when moving an object out at an arm's length. If you pick up a heavy book with both hands and hold it close to your chest you can hold the position without much difficulty. Now hold a book out at full arms' length for a couple minutes and your arms will shake with fatigue.

The further we bend, the further we reach and the higher we reach, the more work we do. These concepts are referred to as ergonomic considerations. A person sitting at a computer should have their shoulders down and elbows close with greater than a 90° angle at the elbow. This means that if you are reaching up to the keyboard *at all* there's constant muscle spasm at the front of the shoulder.

Rather than doing 'active' work with muscle, the body will 'rope up' the arm with fibrous tissue. A little inflammation in the muscles of the shoulder creates this toughened, ropy tissue that helps hold up the arm and may prevent you from feeling the pain. Often, therapy is provided for the top and back of the shoulder but not the front of the shoulder. The shoulder is a shallow joint that allows a large range of movement. The shoulder is stabilized with muscle tissue and is very complex. It takes specialized training and specific deep therapeutic massage techniques to address this presentation.

The inflammation in the shoulder creates pressure that decreases circulation to and from the arm and hand. With less efficient circulation, excess fluid stays in the hand and wrist, which causes them to swell. The secondary swelling, called edema, in the hand then needs to be drained from the hand with massage therapy or lymphatic drainage techniques.

We have briefly described how work posture creates tightness in the shoulder, causing swelling in the hand, which is then diagnosed as carpal tunnel syndrome. Yes, technically this is carpal tunnel syndrome. However, this carpal tunnel syndrome is usually secondary to the shoulder problem we described. The shoulder problem is thoracic outlet syndrome (TOS).

Another cause of this muscle spasm is tension. Americans are good at working harder but not so good at letting go. Some patients are given specific exercises or psychotherapy to help reduce their stress levels. It is interesting to note that dietary factors, such as caffeine and nicotine, also aggravate this problem. However, most people can resolve this shoulder or hand syndrome without dietary changes.

Getting the Best Treatment:

We describe our evaluation and treatment procedures later in this book. As already mentioned, some patients have simply taken our treatment protocol to a massage therapist and had a resolution of their problems, though this is not recommended.

This shoulder-hand presentation is usually chronic and without pain. A person may first become aware of the problem because they try to pick up a cup of coffee and drop it or cannot take the lid off a jar. Though ignoring these problems is easy to do because of the absence of pain, permanent damage may result.

We recommend that you go to a health care provider that has been trained in our technique and is listed on our Web site. Typically most patients resolve in 3 to 17 treatments. As described in our research later in this book, 85% of our patients had a complete resolution of their problem in an average of 3.2

treatments. In situations that are very chronic with some degree of nerve degeneration, patients required an average of 17 treatments. A few patients experienced a decrease in their pain, numbness and tingling of about 60% and decided to stop treatment. Additionally, about 5% of our patients require maintenance care at the rate of one treatment every one to three weeks.

In our perspective, the **most qualified person** to perform our treatment of non-surgical neurovascular decompression is a physical therapist familiar with therapies such as ultrasound, hot pack and other manual therapies. Another excellent choice is a **massage therapist** who has specific training in this technique. A qualified health-care provider, a chiropractor or physical therapist certified in our technique would perform the evaluation and prescribed treatment. Written material would then be provided for the patient and the therapist would perform the manual massage therapy. Massage therapists have successfully duplicated the treatment we describe following participation in one 8-hour training seminar.

Dr. Rahn's Perspective

I have spent considerable time touring the United States from Honolulu, Hawaii, to Clinton, Connecticut, and St. Croix. I worked with a doctor of ergonomics, a true pioneer in the field. We worked mainly with Fortune 500 corporations. This doctor developed simple, easy to use software programs for people to perform their own ergonomic evaluations where they work, saving big corporations a great deal of money when they were fined for unsafe working conditions. She would identify problem areas and document positive changes designed to correct the problems. We worked with both office and industrial settings. One of our goals was to create a medical management program to work hand-in-hand with the ergonomic program.

As a health care provider I can work on someone's hand pain. However, if I do not correct the ergonomic cause of the problem, those symptoms will not resolve. While working on-site, I correlated the ergonomic cause of the problem with the actual medical presentation. Almost without exception, these big corporations reviewed my report and then terminated my relationship with them.

In one large plant I learned that the project leader's boss was related to the company doctor. I was terminated because I was competing with the established doctor. In another large plant I found a supervisor who had worn a rigid wrist brace for 12 years. The supervisor had lost most of the range of motion and muscle strength in her hand. This situation constituted a potential medical malpractice lawsuit because the brace should have been removed after the acute phase of care and the patient should have undergone rehabilitation to improve function.

In brief, I have seen hundreds of patients scheduled for surgery and we have prevented the surgery in most cases with our treatment protocol. But never once did I receive a phone call from the treating doctor asking me what I did to fix that patient. Yes, doctors are busy and, yes, what I do may be unconventional. However, if another doctor cured my patient and I cared about my patients, I would call and ask what they did. Unfortunately, doctors have become business people. There's a saying in India: "When a doctor becomes a businessman you do not have anyone to go to when you're sick."

I am absolutely not a conspiracy theory person. For many, many years I believed that doctors simply wanted their patients to get better. Unfortunately, my experience with corporate workers' compensation doctors is that there is a lack of receptivity to any procedure that competes with their established protocols.

Dr. Rahn's **Thoughts on Chiropractic**

Chiropractic is very powerful. If chiropractic were not effective, it simply would not be as popular as it is today. Sometimes muscle spasm cannot be resolved without some joint mobilization. Conversely, deep therapeutic massage can resolve some joint problems without the traditional chiropractic thrust. For those of you who are new to chiropractic, chiropractors use their hands to locate a joint that has reduced movement. The chiropractor then can make a quick thrust, a short quick movement, to restore joint range of motion. Often, the patient feels an immediate sense a relief, decrease in pain, and improved movement.

There are many different styles of chiropractic care. Some chiropractors only adjust the spine while others only adjust the upper vertebra or two in the spine. Some chiropractors incorporate treatment to the joints outside of the spine. Many chiropractors utilize a full array of physical therapy, nutritional guidance and so on. It is my personal viewpoint that chiropractic should not utilize regional adjustments, which refers to applying the adjustment across the entire neck or low back. Instead, the chiropractor should evaluate the movement for each joint, identify the one joint that has reduced movement in one direction and then apply a specific gentle adjustment for that vertebra. Additionally, the chiropractor should double-check after the adjustment and see if the desired results have been achieved.

Another practice I dislike is applying a forceful adjustment through an area with muscle spasm. I believe that therapeutic massage or physical therapy should be used to reduce the spasm, allowing a gentler adjustment that tends to last longer.

There is a tendency with health care providers to be very territorial. Chiropractors do not want physical therapists to mobilize joints and physical therapists have areas of expertise that they feel belong to them and so on. I know that this treatment protocol will aggravate some of these traditionalists. However, the main consideration is the patient's well being.

What makes our technique so very effective is that we combine chiropractic, physical therapy, and massage therapy. I think of this combination of therapies as being like a 'chef's secret recipe'. The recipe may be simple and easy; however, nobody but the chef knows it. Well, I don't want to be the chef with the secret. I believe we need cooperation between disciplines and we need health care providers to do whatever we can to help the patient to the best of our ability.

I also believe that providing health care is an honor and a privilege and that a health care provider's concern for making a profit should be secondary to taking care of the patient. For me, my work is a compromise between profit and effectiveness. With longer, more effective treatments, I cannot see as many patients in a day and therefore make less money than many chiropractors. We have found that treating carpal tunnel syndrome with this technique requires about 30 minutes of care. Probably, the solution to the problem is for chiropractors to hire massage therapists to do the actual soft tissue techniques. The medical doctor, chiropractor, or physical therapist can do the initial evaluation and follow-up evaluations while leaving the massage therapist or physical therapist to do the bulk of the hands-on work.

Physical therapy is powerful. Chiropractic is extremely effective.

Massage therapy is essential for this type of treatment. Combine all three and the patients get better quickly with the smallest cost. Personally, I do not feel it is wrong for a patient to take pain or anti-inflammatory medications. However, because all medicines have side effects and, for this type of problem, do not cure the cause of the problem, medication should be provided in the acute phase of care but not long term.

My viewpoints are different than those of other health care providers who **are** qualified and effective, but they are offered respectfully and with the hope of increasing cooperation between different types of health care providers.

Why Most Carpal Tunnel Syndrome Diagnoses Are Inaccurate ***(this section is written primarily for health care providers)***

Research has proven that pressure on a nerve can cause a myriad of symptoms. Three examples of this are: disc herniation (20), bony encroachment on the nerve root (21), and pressure below the spinal cord, which causes sciatica (22). All of these presentations can put pressure on the nerve root and cause pain, numbness and weakness.

Pressure on a nerve at the spine may be referred to as central compression. Pressure on a nerve in the shoulder or arm may be referred to as peripheral compression or peripheral nerve entrapment. It is possible to have both central and peripheral problems occurring at the same time.

Carpal tunnel syndrome (CTS) is usually defined as pressure on the median nerve at the wrist (10, 11, 12). Medical literature usually looks at 3 types of CTS: Idiopathic 'just happens' with no known cause. Intrinsic is caused by increased volume in the carpal tunnel, putting pressure on the nerve. Extrinsic has to do with the carpal bones such as subluxations, fractures and arthritic changes (11).

Conventional diagnostic tools (including X-ray, MRI, EMG and nerve conduction studies) have certain critical limitations. While X-ray and MRI are commonly used to indicate pressure on a nerve at the spinal level (spinal cord or nerve root pressure from bone or disc problems), **these** diagnostic techniques are not effective in diagnosing pressure on nerves from muscle or soft tissue. EMG and nerve conduction studies indicate nerve disease and pathology but usually do not pinpoint the exact location where the problem started. Or, nerve pathology may be found at the wrist but the original cause of the edema is the shoulder. Although the public tends to feel that radiological findings of a disc bulge 'prove' the cause of their pain, about one-third of the population under the age of 30 has asymptomatic disc bulges (23, 24).

Other orthopedic tests used to diagnose CTS include Tinel's, Phalen's and hand elevation (13). However, this hand presentation may still originate in the shoulder. Research has also demonstrated that the volume of the carpal tunnel and the size of the median nerve may increase after a surgical release (8). This carpal tunnel 'bottleneck' may also be caused by hand edema (swelling) secondary to pressure on the vessels at the front of the shoulder. Other effects on the normal movement of material down the nerve have been published by Dr. Rahn (1).

Our treatment protocol utilizes palpation, in which the doctor uses his or her hands to feel for muscle spasm, and Adson's test to determine if there is pressure on the blood vessels and nerves at the front of the shoulder. Ironically, these simple tests are quicker and less expensive than traditional radiological and nerve conduction examinations. These procedures are described in more detail later in our 'Treatment' chapter.

As mentioned earlier, typical drug intervention might include the administration of anti-inflammatory medication in addition to pain medication. Anti-inflammatory steroids are effective in short-term treatment of CTS (25, 26). Chronic severe presentations might be treated with surgical release, a surgical neurovascular decompression in which the ligament at the base of the palm is cut. This procedure is reportedly 90% successful (27, 28, 29, 30).

However (28): There is a long time (9.8 months) to maximal improvement. Thirty percent of patients report poor to fair strength and long-term scar discomfort and 57% of patients report a return of pre-operative pain beginning 2 years after surgery. Also, patients that had surgery 3 or more years after the initial diagnosis were less than half as likely to have symptom resolution (29)! Two additional consequences of the surgery are fibrous tissue (scar tissue itself puts pressure on the median nerve) and decreased stability of the ligaments. This post surgical 'compromised wrist integrity' often requires the patient to undergo costly vocational rehabilitation. To summarize, conventional treatment utilizes relatively ineffective techniques to postpone surgery, even though surgery is less likely to be successful the longer a patient waits after initial diagnosis.

We contend that the true cause of the problem, stemming from the shoulder, has not been identified or treated. First, muscle spasm at the front of the shoulder (primarily pectoralis minor) creates pressure on the nerves that run to the arm and hand (brachial plexus). This can have the same effect as pressure on a nerve root at the spinal level observed with disc herniation. This muscle spasm also creates pressure on blood vessels. Decreased circulation can cause swelling (edema) in the hand as previously described.

Now we come to the crux of the problem. If doctors hear the patient complain of pain and numbness in the hand, detect swelling in the wrist and perform an EMG that shows nerve pathology, they feel confident with the diagnosis of CTS. Technically, this diagnosis is correct. However, this problem usually does not originate in the wrist. Muscle spasm at the front of the shoulder created both swelling in the hand and pressure on the nerves. This diagnosis is called **thoracic outlet syndrome** and, as mentioned, this condition is virtually undetectable with X-ray and MRI.

Very little clinical evidence exists to demonstrate the effectiveness of chiropractic and massage therapy in treating CTS. One paper suggests chiropractic care is effective, but the study lacked sample size (18). Another paper determined that chiropractic care is as effective as the use of ibuprofen when median nerve degeneration has not progressed from demyelination (31) to axonal degeneration (19). A review of 43,000 workers' compensation claims over a period of 19 years found that chiropractic care is less expensive and more effective than conventional medical care (34). The review found basically the same treatment cost per patient as found in our research described later in this book (34). What is unique in our study is the combination of physical therapy, joint mobilization, and massage therapy.

Making a Correct Diagnosis

Many times carpal tunnel syndrome (CTS) and thoracic outlet syndrome (TOS) have no pain, and the only symptom that presents is weakness when opening a jar or, for example, when you drop a cup.

Pain, weakness, and numbness in the hands usually indicate a problem with the nerves. However, this does not determine if the problem is from the neck, shoulder or hand! Swelling in the hand may be from muscle spasm in the shoulder, arm, or hand. In short, you **cannot diagnose carpal tunnel by the presence of pain, numbness or weakness in the hand** because these problems can come from the neck, shoulder, arm, or hand. Again, an X-ray usually will not diagnose muscle spasm, and an MRI that shows a disc bulge in the neck does not mean that the bulge is causing a problem. Many people have bulges without any symptoms.

We developed a **symptom survey to allow anyone to determine if his or her hand symptoms come from a problem in the neck, shoulder, or hand.** It does not constitute, and should not replace, a proper diagnosis by a health-care provider. The symptom survey is simply meant to educate the public as to what different areas of the body may cause hand problems. Since more than one area may contribute to the problem, complete all sections of the survey.

The discussion thus far focuses on CTS. However, similar clinical presentations include sciatica (pressure on the sciatic nerve from the pyriformis muscle) and quadriceps weakness from pressure due to muscle spasm and lymph edema on the femoral nerve (anterior femoral nerve entrapment syndrome).

Survey for the Neck:

- Does the pain seem to come from the neck (the center or midline where the spine is)? If yes, give yourself 10 points.
- Is the pain aggravated or does the pain occur: i) on the side opposite to the direction you move as you turn your head to the side, ii) when you bend your head forward, or iii) move your ear toward your shoulder while looking straight ahead? This pain may be from muscle spasm. If yes, give yourself 5 points.
- If someone else pushes gently on the top of your head while you are sitting, does the pain become sharp and shoot out from the middle of the neck to the

shoulder or arm? If yes, give yourself 15 points.

- Have someone gently massage the tops of your shoulders while you lay on your back. If this reproduces your pain, subtract 5 points from your score.

If your neck score is over 20 points it is likely that you have a neck (spine) problem, which is usually pressure on a nerve root or a disc problem. This score warrants an exam by a health care provider because, if left untreated, you may suffer a loss of strength, sensation and range of motion. Permanent disability can result.

A score less than 20 points may simply be from muscle spasm and a massage therapist can take care of the problem. However, any persistent problem should be evaluated by the appropriate health care provider.

Survey for the shoulder:

- Press below the collarbone, move your fingertip from left to right, and feel from the center of the chest to the armpit. If you feel hard bumps (muscle spasm) and tenderness, give yourself 10 points. A health care provider would be looking for the muscle named 'pectoralis minor.'

- Feel the pulse in your wrist with the index finger of the other hand (on the inside of the wrist toward the thumb side of the arm). If it is weak give yourself 10 points.
- If someone else can feel your wrist pulse while gently pulling the arm to the side, but the pulse strength *changes,* diminishes or disappears as they pull it back, give yourself 15 points.
- If the muscles around the shoulder are tender when pressed, or the shoulder is painful when moved, give yourself 10 points.

If your shoulder score is 20 points or higher, there is a good chance that you have thoracic outlet syndrome (TOS). It compresses the nerves and blood vessels in the front of the shoulder, which may cause swelling in the arm and hand. TOS can cause weakness, numbness, and/or pain in the hands.

A score less than 20 points may simply be from muscle spasm, and a massage therapist can take care of the problem. Again, any persistent problem should be evaluated by the appropriate health care provider.

Survey for the Hand (Carpal Tunnel Syndrome):

- Place your hand flat with the palm down on the table. Press firmly on the high point of the wrist. To find this point, go to the midline of the arm, find where the wrist joint bends, and move 1 inch towards the fingers. If the wrist ***bones*** feel springy (move more than 1/8 inch) you probably have carpal tunnel syndrome! Dr. Rahn's wrist brace and treatment protocol is definitely indicated.

Summary

At our clinic and the clinics of those trained in our treatment protocol, we determine the true causes of the problem and treat each accordingly. Often, there are contributing factors from **more than one area.** If you have TOS, the shoulder must be treated. If you have CTS, you should use Dr. Rahn's brace instead of the traditional brace that pushes the wrist up and back, thereby increasing the pressure on the carpal tunnel.

Also, it is okay to wear a brace to limit movement, reduce swelling through compression, and reduce pain

regardless of where the problem originates! Just know that a brace is for **temporary use**, as prolonged use decreases range of motion and strength. You should use Dr. Rahn's 'full hand' compression glove if there is swelling in the fingers.

You can read more in the written treatment protocol in the next chapter. You may also obtain a DVD of Dr. Rahn teaching this protocol for your own information. Some patients have found a local massage therapist, shared the DVD with the therapist, and successfully treated their pain and disability.

Dr. Rahn's Web site lists health care providers that have taken his workshop and successfully passed the course. Please feel free to call if we can be of further assistance.

Treatment Protocol for Non-Surgical Neurovascular Decompression

The non-surgical neurovascular decompression (NSNVD) treatment protocol is guaranteed to produce a 50% reduction of pain within 5 weeks. It is a compilation of existing chiropractic, physical therapy, and massage treatment methods used to relieve the compression that causes carpal tunnel syndrome (CTS) and thoracic outlet syndrome (TOS) from non-bony, such as large first rib, causes. This protocol is performed in a unique combination and utilizes a patent-pending carpal tunnel syndrome decompression glove or taping.

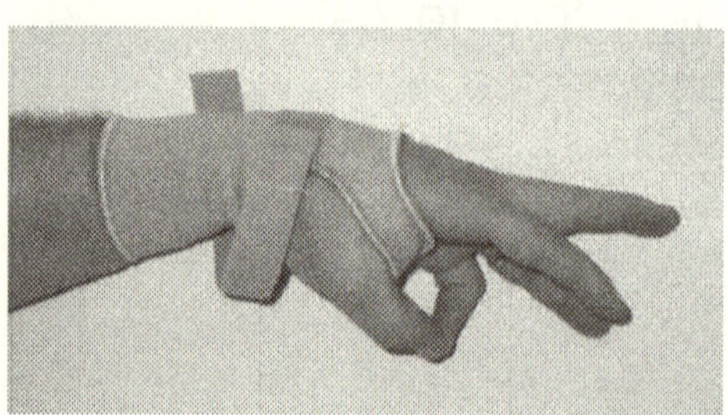

This section is addressed primarily to patients without medical training, but a format for health care providers is available upon request. The goal is to address improvements in conventional treatment and raise awareness of aspects of treatment that should be included in the overall treatment regimen. References to medical terminology are placed in parenthesis. This treatment must be administered by a health care professional. Patients must be examined and evaluated by a medical doctor, doctor of chiropractic, or physical therapist to determine the cause, complications, and co-existing health issues and then determine a specific treatment for each individual. Patients that self-administer this treatment without appropriate medical training or licensing may misdiagnose themselves and cause damage.

This chapter presents a simple description of a complex set of symptoms and the components of treatment utilized by NSNVD. The NSNVD protocol utilizes variations of the following diagnostic and therapeutic methods: differential diagnosis, ice, ultrasound, massage, mobilization, brace, rehabilitation, ergonomics and education.

Differential Diagnosis:

Pain, numbness and tingling in the forearm and hand can come from insufficiency or compression of the blood vessels and nerves in the neck, shoulder, elbow, and wrist. Less common causes include diseases such as diabetic and cardiac involvement. Other exam findings include a decrease in light touch, muscle weakness (decreased grip strength and pinch strength), and abnormal deep tendon reflexes. A practitioner can obtain a good history, look at the patient's subjective complaints and then perform cervical spine compression and spinous percussion to determine if a herniated disc is causing the hand pain. Remember, a positive MRI for disc bulge is not proof positive that the disc is causing the pain! The diagnostic process must not stop with positive X-ray or MRI findings.

Muscle spasm in the front of the shoulder compresses nerves and vessels causing wrist swelling as a secondary problem, CTS. Feel your wrist pulse strength. Then, have someone pull the arm back and see if the pulse is weaker. This is called a positive Adson's sign and indicates a cause for hand and wrist swelling from shoulder spasm. A practitioner can properly evaluate this and determine a diagnosis of thoracic outlet syndrome (TOS). Treatment should then include the shoulder.

Traditionally, TOS was thought of as a large first rib (bony) impingement on the nerves and blood vessels of the neck. TOS from muscle spasm is rarely given proper consideration.

Determining compression of nerves in the forearm is a process of feeling (palpating) for areas of tenderness, muscle spasm and heat from swelling. Orthopedic testing, muscle testing and range of motion also help to complete the determination of what structures are affected. As mentioned above, a TOS problem can create secondary presentations of nerve compression in the arm or hand such as CTS. If there is pain when you move, the problem may be from muscle-tendon involvement. If passive movement (the tester moving the joint) results in pain, the problem is usually due to ligaments or joint inflammation.

Of course, there are other commonly used orthopedic tests to make the diagnosis of CTS. For example, practitioners may use X-rays or tuning fork testing, which finds pinpoint tenderness, to rule out fractures.

What we have focused on in this process is determining if the CTS came from a soft tissue problem in the shoulder (TOS), or what proportion of the presentation is from the wrist rather than the shoulder.

Note to practitioners: Thenar atrophy from other causes (disuse or other neuropathies) and pain due to osteoarthritis of the first carpal-metacarpal joint are sometimes confused with CTS. Trigger fingers and deQuervain's stenosing tenosynovitis are commonly associated with CTS.

Ice and Ultrasound:

Patients frequently use heat for pain. This may feel good initially but usually increases swelling, which will cause more pain for the next few days! Cold tightens the tissues, reduces swelling, decreases muscle spasm, and reduces pain. Ice or a flexible ice pack wrapped in a moist hand towel placed on the affected area (front of the shoulder and wrist) for 10 minutes is sufficient. Another method of icing is to put ice water in the sink or a large salad bowl and immerse the hands to a level above the wrists for 8 minutes.

The elderly and people with systemic nerve problems, such as those with diabetes, need to exercise caution as frostbite can occur. To transfer the cold effectively, a moist towel should be wrapped around the ice pack. Do not ice more than once every 2 hours.

A good health care provider will direct you to ice areas that contribute to the problem even if they do not hurt. An example is icing the front of the shoulder for TOS when the nerve pain and pain from swelling is in the wrist. Again, relieving the spasm in the shoulder will allow the hand to drain and thus alleviate pain.

Other ways to reduce swelling and pain include treatment with ultrasound and ultrasound with muscle stimulation. In addition, ultrasound may help treat the shoulder tightness (include the deltoid, capsule and rotators in your treatment), which can put pressure on the nerves and vessels going into the arm and ultimately cause the CTS swelling in the hand.

Massage:

Deep therapeutic massage to tight muscles in the upper trunk, arm, and neck is essential for most patients. It reduces the spasm and also promotes circulation, thereby draining the swelling. A massage technician trained only in relaxation or Swedish massage is not qualified to perform these techniques. A 'Trigger Point' or 'Shiatsu' type of treatment is needed for the muscle spasm from the hand all the way up through the shoulder and neck.

If scar tissue is detected, another specialized technique is also essential. This scar tissue technique should have the soft tissues shortened (approximated) so that muscle tissue is not damaged. Then, a technique called effleurage, which uses light, gliding movements from the extremities towards the trunk, is needed to reduce swelling (edema) in the entire arm and shoulder.

Lastly, if there is a problem with deep lymphatic drainage, such as when the lymph nodes have been surgically removed, a specialized superficial lymphatic drainage technique must be employed. This technique may use specific exercise, bandaging, or a non-elastic sleeve to optimize results.

<u>Mobilization:</u>

A joint mobility evaluation and treatment should be provided as needed to the neck, shoulder, upper rib cage (anterior thoracic vertebra), and wrist, especially the carpal bones. Joint mobilization reduces joint pressure; which then has the effect (through the nerves) of reducing pain and muscle spasm.

Chiropractors may use a 'short amplitude high speed thrust' through the 'end' of the joint range of motion. Physical therapists, massage therapists, and other practitioners may oscillate the joint or use other techniques to achieve similar results. Joint mobilization may relieve muscle spasm, though some muscle work before the mobilization usually helps the mobilization technique.

Brace:

A brace serves two functions: First, if the ligaments are stretched, it stabilizes the wrist and allows healing. Second, a full glove may be used to reduce edema, as will compression gloves or stockings.

This is another instance in which our protocol differs from conventional treatment. A typical carpal tunnel brace limits your range of motion with either cloth bindings or a rigid stay (bar) and may position the hand up in extension toward the back of the forearm. This position compresses and aggravates the hand, wrist, and forearm. It is believed that wrist extension compresses the carpal tunnel in a wrist with **damaged, over-stretched ligaments**. We are conducting clinical studies to determine whether the wrist extension position compresses the contents of the carpal tunnel, thereby aggravating the condition.

Our technique uses a glove with a 'hook and loop' (Velcro) fastener from the thumb side of the base of the palm to the little finger side of the base of the palm. It is important to touch the tip of the little finger to the thumb tip

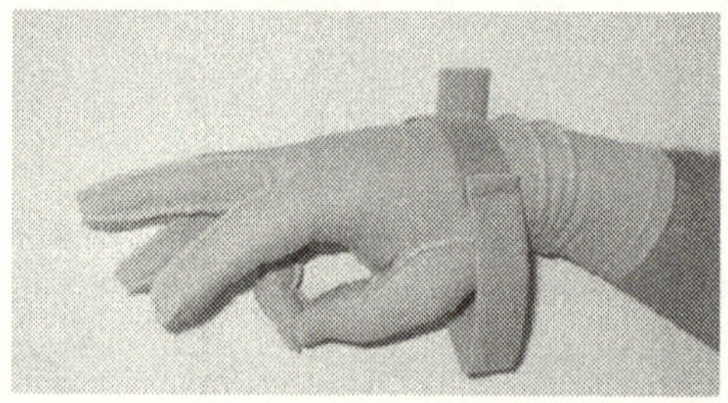

and flex the wrist (tip the hand toward the inside of the forearm) a little before fastening the strap. It is believed that this position will push the top of the carpal tunnel up and out, decreasing the pressure on the tunnel contents. In our research and practice, we use a caliper to measure this tunnel 'height' in wrists with ligamentous (CTS type) damage.

If this type of brace is not available, a practitioner can place the hand in the same position and wrap one and a half turns of 1-inch athletic tape around the wrist on the base of the hand without taping over the joint between the forearm and the hand. See the illustration of the patent-pending 'neurovascular compression glove' at the end of this section. The glove need not have the fingers covered.

Note on the Surgery Alternative: We highly recommend that patients have at least a trial of conservative treatment, like that described in this book, before surgery. With surgery, the chances of spontaneous healing are often significantly reduced.

Full decompression glove and open wrist brace, both with wrist strap:

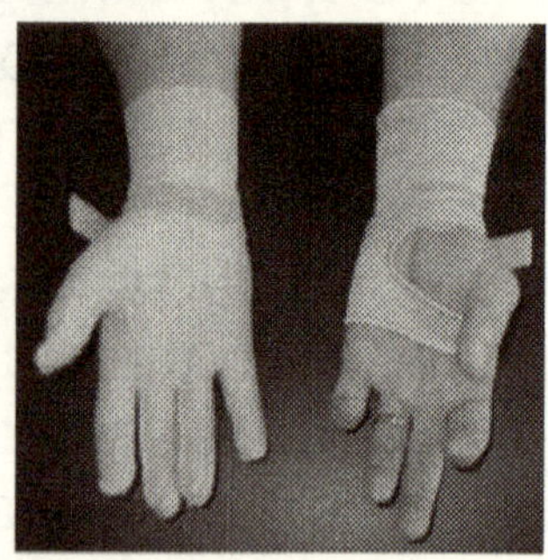

Rehabilitation:

Please see Appendix J for the typical exercise and stretching program we provide for patients. This format was printed from ToolsRG Software that is copyrighted by PhysioTools, Limited, and can be found at: www.ToolsRG.com.

It is important to evaluate whether the wrist is stable or has ligamentous damage. If the ligaments are damaged, the glove must be used to prevent stretching of the wrist (flexor retinaculum) during the stretching and exercising. A practitioner may place your hand on a hard surface and press on the high point of the carpal tunnel to check for the springy feel of damaged ligaments. Experience will help determine if there is excessive movement, indicating ligamentous laxity. Usually, there is **very** little movement, called joint play, with movement of less than 1/8th of an inch. If the wrist 'flexor retinaculum' is unstable or overstretched, pressing on the high point of the carpal tunnel will cause the tunnel to 'flatten out' since it is the base of the tunnel that maintains the curve.

In the initial, acute phase of treatment, rehabilitation should include home icing for eight minutes once a day and 24-hour-a-day use of the glove or taping. If you lean your bodyweight onto the wrist while getting up at night to go to the bathroom, the ligamentous integrity will not improve. Please note that while most practitioners believe a stretched ligament will never recover, a significant degree of stabilization is achieved if the wrist is in the proper position as illustrated below (rounded as when the thumb and first finger approach each other) for at least six weeks.

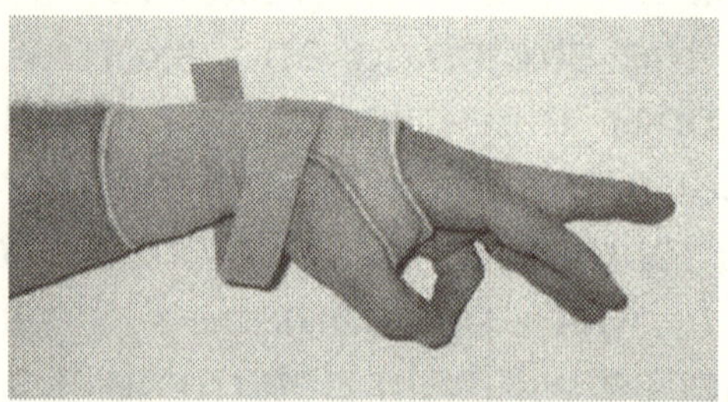

Ergonomics and Education:

Many CTS presentations are due to repetitive stress caused primarily by your position and posture. The wrist is tipped up too far during typing, the mouse is too far away such that the weight of the arm causes TOS, and so on. If you want CTS to clear up and not return, have a proper ergonomic evaluation of the workplace as well as the home!

Reconsider carrying children, moving heavy groceries into the trunk, reaching 'high' in the kitchen and so on. Low or no cost ergonomic changes (such as placing a ream of paper under the computer monitor to raise it up) must be made in the workplace and home.

If you unconsciously hold tension in the muscles of the neck, shoulders and arms, further education must take place. At times, stress reduction techniques, including biofeedback, guided imagery, and other exercises, are shared to complete the health care regimen.

The most important factors in preventing a re-occurrence of the problem after symptoms have resolved are to **continue** as a lifestyle commitment with: 1) an exercise and stretching program, 2) stress reduction (refer to Appendix B), and 3) maintain good posture and ergonomics. Appendix A discusses ergonomic considerations in more depth.

Closing comments:

Pain sufferers may feel free to e-mail Dr. Rahn for dialog on this topic. There is no substitute for the full program under providers properly trained in these techniques. Seminars in this technique are available as are a training DVD and booklet to supplement the seminars.

This non-surgical neurovascular decompression protocol may be applied to other similar ‘neuro-musculo-skeletal’ presentations as described in Appendix D (regarding neck pain, TMJ and headache), and Appendix E regarding leg pain and weakness. Evaluation and treatment includes considerations in joint mobilization (either chiropractic manipulative therapy with a 'thrust' or 'non-thrust' grade 3 and 4 mobilizations), therapeutic massage (fibrous tissue reduction, Trigger Point types of techniques, venous circulatory, and superficial lymphatic), ergonomic considerations, exercise and stretching, decompression braces, physical therapy modalities (especially ultrasound with or without electric muscle stimulation) and self care at home. Other holistic considerations include meditation, stress reduction, fasting, and dietary and supplement considerations. See Appendix C and Appendix B, respectively, for a discussion of nutrition and stress reduction.

Satisfaction Guarantee

We are so confident of being able to reduce chronic pain that we give our patients a satisfaction guarantee. The guarantee is that the patient shall have a 50% reduction in chronic pain within five weeks of beginning treatment or the portion of payment contributed by the patient will be returned.

This agreement makes clear that the patient must also work to achieve the desired results. The patient agrees to:

1) Do whatever daily stretching, icing and exercising the doctor requires (this is called 'home care').
2) Not miss any scheduled appointments.
3) Practice a daily 15-minute 'stress reduction' technique if prescribed by the doctor.

If the patient is initially taking painkillers or other medication on a daily basis, we estimate the baseline pain level as if the patient is not taking any medications.

If the problem involves decompression and regeneration of a long nerve (such as from the neck to the hand or the sciatic nerve in the leg) the process can take two to three months. In this situation, even if the pain is not reduced by 50% in five weeks, the treatment may be considered successful if there is a 50% improvement in objective findings (return of light-touch sensation, return of muscle strength, increased range of motion, reduced chronic muscle spasm, and so on).

An exception to this guarantee is where nerves have degenerated to a degree that regeneration is unlikely. In this rare instance, the doctor will evaluate and inform the patient in advance of any treatment.

Research Supporting Our Treatment Protocol *(this section is written primarily for health care providers)*

We conducted a clinical trial using our treatment protocol with 100 patients. The study was conducted in a single center, our office, over a period of about three years, from November 2002 to August 2005. A quick reference abstract has been provided to summarize the methods and results.

Abstract:

One hundred patients were treated (**155** peripheral nerve entrapment presentations and compression of the nerves outside the spinal cord) including: 85 with hand symptoms, 14 with sciatica and one with femoral nerve compression. Physical examination determined if hand pain originated in the spine, shoulder, or hand without costly radiological testing.

Eighty-five of the patients completely resolved. Of these, 86% required an average of 3.2 treatments and 14% required an average of 16.5 treatments. Five patients require maintenance care at one treatment every one to three weeks. Ten patients are permanent and stationary (maximally improved). **The fifteen patients who did not experience full resolution rated their reduction in symptoms at about 60%.**

Half of all patients have been tracked for more than one year. Only 10% had mild flair ups within two to six months, which was successfully resolved with an average of two treatments.

All patients presenting with hand symptoms had the primary cause of their presentation in the shoulder.

Treatments took 30 to 45 minutes. The cost per treatment was $52 (at California Worker's Compensation rates). 90% of the treatment time was focused on the shoulder. The two groups treated averaged $166.40 and $858.00 per patient to completely resolve or reach maximally improved. Reoccurrence and comparative cost analysis is included.

This highly effective, inexpensive treatment incorporates massage, physical therapy, and joint manipulation (chiropractic or non-thrust). Experienced massage therapists successfully duplicated treatment after one 8-hour seminar.

Methods:

We treated 100 patients between November 2002 and August 2005 (two years and nine months) following the treatment protocol beginning on page 42.

To summarize, treatment consisted of the following on all patients:

- Deep therapeutic massage (DTM) to all of the shoulder muscles (primarily pectoralis minor and anterior scalene);
- Chiropractic manipulative therapy (CMT) or non-thrust joint mobilization to the shoulder joint (glenohumeral joint) and wrist (carpals) as needed;
- A stripping massage or lymphatic technique to clear edema; and
- 'BioFreeze' analgesic balm to the spasmodic muscle contractions (myospasm).

Acute patients received ultrasound and ice, whereas sub-acute patients could receive hot packs to the shoulder (anterior and posterior). Home care consisted of ice packs and stretching in the acute phase and strengthening in the chronic phase of treatment (see Appendix J). We utilized a Symptom Survey on all patients treated, (beginning on page 32). These patients suffered from the following peripheral nerve entrapment syndromes:

- 85 with TOS and hand symptoms (some also had sciatica),
- 14 with sciatica, and
- 1 with femoral compression.

Though 100 patients were treated, 155 syndromes were treated because some patients had bilateral hand involvement or hand and sciatic presentations. Five patients received a more detailed examination using the 6-page form described in Appendix F.

Results:

Out of 100 patients treated, 85% of the cases resolved (the patients felt they no longer had any symptoms and there were no objective findings). Subjective data was gathered in which the patients described their pain as follows:

- Minor (0-33%)
- Low to Moderate (33-66%)
- Moderate to High (66-100%)

So a drop from 'Moderate to High' to 'Low to Moderate' would constitute a 33% drop in symptomotology. Initially, 65% of patients were subjectively and objectively rated at a severity of 'acute'. Six percent were rated as mild, 14% as low to moderate, and 15% as moderate.

At the end of the study, five patients continued to require maintenance care at one treatment every one to three weeks. This 5% of the patient population rated their subjective improvement at an average 59%. It is the opinion of Dr. Rahn that long-term work on ergonomics (evaluating workstations and home issues), stress reduction, and diet can eventually resolve almost all patient presentations. Of course, lifestyle changes take time and require receptivity on the patient's part. Ten patients decided to stop treatment or are 'permanent and stationary' (maximally improved) without requiring maintenance care. This 10% of the patient population rated their average improvement at 63%.

We categorized the patients by the number of total treatments before resolving their presentation or reaching the 'permanent and stationary' status. The group requiring a large number of treatments is 14% of the total population and they received an average of 16.5 treatments. The group requiring a smaller amount of treatment is 86% of the total population and this group received an average of 3.2 treatments.

This study included 14 patients with peripheral nerve entrapment as "Pyriformis Syndrome" putting pressure on the sciatic nerve. This '**sciatic**' presentation has the same profile and considerations as do the rest of the patients with TOS (hand pain from thoracic outlet syndrome). One patient had peripheral compression of the **femoral nerve** (anterior femoral nerve entrapment syndrome) and, again, the overall treatment presentation was consistent with the TOS patients. The only difference with these other two diagnoses is in how the treatment is applied to a different part of the anatomy, which is described in Appendices D and E. These other presentations are also discussed on our Web site and DVD.

Interestingly, many of these patients continued treatment over a period of many months **after** their presentation resolved because they had some other problem, such as neck stiffness, or wanted maintenance treatment. This is an important point because we have determined that **90% of the 'hand pain' patients, once resolved, do not have a recurrence of their hand symptoms**. About 5 to 10% of the patients wanted infrequent maintenance care for soreness or stiffness, which prevented a reoccurrence of the hand symptoms. Rarely, if a true flair up reoccurred, which happened once in two to six months, resolution was achieved with an average of two treatments.

<u>Special considerations for the five detailed hand-examination patients:</u>

None of the five patients measured exhibit a change in the carpal tunnel height from wrist flexion to extension. This supports the hypothesis that these patients had hand symptoms secondary to TOS. Although research has been conducted regarding medical imaging of carpal tunnel volume (14, 15), no research has been conducted to differentiate CTS patients into those with and those without ligamentous damage and then measure the change in volume with positional changes in the group with ligamentous damage. The expected finding is that there would be a significant reduction in carpal tunnel height (measuring palmer to dorsal, or anterior to posterior), which would correlate to an increase in pressure on the median nerve (and subsequent pathology consistent with CTS).

One patient, '97', had a resolution in hand symptoms without any change in exam measurements. It is felt that this patient's presentation was short-term and caused by excessive weight lifting. The subjective symptoms and objective findings (palpable TOS myospasm and a positive Adson's test) resolved. Two other patients, '33' and '63', showed dramatic increases in strength as follows:

<u>Patient "63":</u>

Increased average grip strength

left 18% right 6.4%

Increased sustained grip strength

left 20% right 15%

Increased Pinch Strength

Thumb to second (digit)
left 50% right 18.4%
Thumb to third

left 38.3% right 28.9%

Thumb to fourth

left 23.1% right 31.9%

Thumb to fifth

left 29.3% right 35.6%

Lateral average increase

left 19.2% right 13.2%

Volume hand/forearm (decrease)

left 18.4% right 02.1%

Patient '33':

Increased average grip strength

left 7.6% right 13.6%

Increased sustained grip strength

left 39.5% right 41%

Increased Pinch Strength

Thumb to second (digit)

left 22.6% right 5.5%

Thumb to third

left 32.7% right 13.2%

Thumb to fourth

left 14.6% right 40.5%

Thumb to fifth

left 27.3%   right 0.0%

Lateral average increase
left 20.0%   right 13.6%

Volume hand/forearm (decrease)

left 5.2%   right 0.9 %

The remaining two patients in this sub-group elected to stop treatment as 'permanent and stationary' before the follow up 'detailed exam' could be obtained.

Discussion:

All 85 patients with thoracic outlet syndrome (TOS) presented with hand symptoms. Many had been previously diagnosed with CTS and several had received carpal tunnel release surgery that ultimately failed. Not one of these patients had true CTS; they all had CTS secondary to TOS. Please note that CTS may exist without TOS, as with a hard fall onto a hard surface in which the flexor retinaculum is stretched at impact (extrinsic CTS). Another true CTS presentation may be from keyboard or industrial work (repetitive stress). Two patients in this group '16' and '73' were in the chronic, high treatment group and both of these patients were chronically aggravated from constant keyboard work. However, to summarize, **most patients diagnosed with CTS actually had a hand problem secondary to TOS** (or, at least, the two diagnoses present concurrently). We also found that **TOS can cause dependent edema**, which seems to contribute to the occurrence of various tendonitis presentations.

This treatment protocol is very effective for the treatment of the TOS/CTS presentations. In summary, we had 85% of all patients resolve their problem, 5% continue to receive minimal maintenance care at one treatment every one to three weeks and rate their improvement at 59%), and the remaining patients discontinued treatment as 'permanent and stationary' without further care and a 63% improvement in their symptoms. Fourteen percent of the 'resolved' patients required an average of 16.5 treatments to resolve their presentation, while 86% of the patients required an average of only 3.2 treatments for symptom resolution. Experienced massage therapists receiving one 8-hour seminar were able to successfully reproduce treatment results in patients.

This timeline is segmented where certain groups resolved their care. Seven patients resolved 'initially' with only one treatment. Eighty-five patients resolved at an **average** of 3.2 treatments. Ten additional patients resolved before their 17th treatment and five patients continued to require maintenance care at the end of the study.

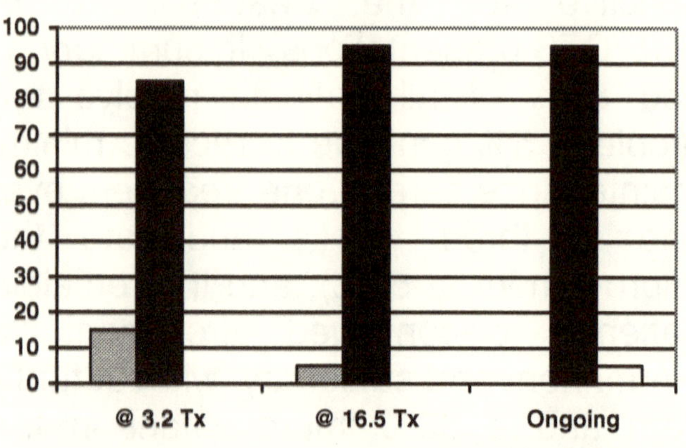

Notably, our treatment protocol treated **peripheral nerve entrapment syndrome equally well whether the**

diagnosis was TOS, sciatica or femoral nerve entrapment. One of our maintenance patients, '100', has been diagnosed with a significant disc herniation and was recommended for surgery. Again, while this treatment is for peripheral nerve compression, a central compression presentation was improved by relieving **concurrent** peripheral compression (a 50% reduction in pain).

A larger sample size of patients is needed to fully evaluate the 'detailed shoulder/hand' examination findings. There is a need to study exactly which nerves and muscles are involved in different presentations. It is likely that standard rehabilitation exercises do not isolate the affected digits or part of the hand. Two of the five patients initially examined with the detailed exam were not re-examined because they discontinued treatment as 'permanent and stationary'. However, these patients initially **exhibited a 4% and 6% change in the carpal tunnel height** with position change (moving into wrist extension). These two patients constitute a new category of 'true' extrinsic carpal tunnel syndrome (ligamentous damage from trauma most likely). And, this presentation should respond well to the use of Dr. Rahn's brace to increase carpal tunnel height (decreasing pressure on the median nerve).

Further Study Needed:

Because the 2 patients with ligamentous damage were not re-examined; and, therefore not considered in this conclusion, the remaining 3 patients had hand symptoms secondary to TOS (shoulder) involvement.

In our years of clinical experience, we have found that the exam findings for extrinsic or traumatic CTS resulting in ligamentous laxity are distinct. For example, Dr. Rahn's hands have ligamentous damage at the flexor retinaculum and a decrease in carpal tunnel height (measuring palmer to dorsal, or anterior to posterior) while moving from flexion to extension by an average of 22% as measured with a micrometer (see Appendix H). We have observed and successfully addressed this presentation throughout years of clinical experience, but we have not documented it.

Although research has been conducted regarding medical imaging of carpal tunnel volume (14, 15), no research has been conducted to differentiate CTS patients into those with and those without ligamentous damage. A study should measure the change in volume with positional changes in the group with ligamentous damage. The expected finding is that there would be a significant reduction in carpal tunnel height (measuring palmer to dorsal), which would correlate to an increase in pressure on the median nerve and subsequent pathology consistent with CTS.

Furthermore, the study should treat the patient group that exhibits ligamentous laxity by **utilizing the glove** (with strap) and **evaluate an improvement in ligamentous integrity over time**. In our study, the glove was used successfully by four patients to compress the hand and decrease edema and pain. Additionally, the glove with strap (to maintain carpal tunnel height while ligamentous healing occurs) should be helpful in patients post-surgically (with a carpal tunnel release).

<u>Cost effectiveness, severity of presentations and work days missed</u>:

The treatment cost per year is compared in the graph on the next page. The long term and maintenance costs were calculated based on the average patient, who would receive 16.5 treatments in the first three months. For the remaining nine months of the year, they would receive an average of two treatments per month for nine months, totaling 18 treatments. The initial 16.5 treatments plus 18 additional treatments totals 35 treatments per year, or $1,820.00 per year.

Treatment Cost in Dollars Per Year

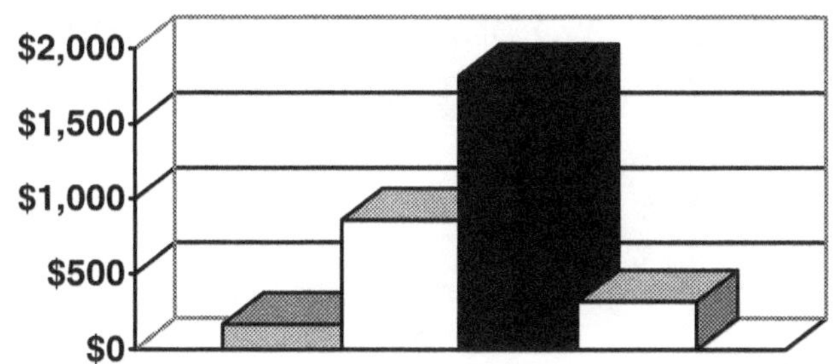

- (1st column) Cost Short Term Patients (3.2 Txs) =$166.40
- (2nd column) Cost Long Term Patients (16.5 Txs) = $858.00
- (3rd column) Cost Long Term + Maint. (1 year) = $1820.00
- (4th column) Ave Cost / Yr For All Patients = $318.12

By comparison, the medical costs and loss of productivity due to CTS add up to $29,000 per patient, according to a report by physicians at the Harvard Medical School (32). The same report estimated that musculo-skeletal injuries in the United States alone total $20 billion per year (32). Another review estimated that half of all Americans will have occupational injuries by the year 2000 (33).

The cost of treatment in our study also compares favorably to typical workers' compensation claim costs as illustrated in the graph below.

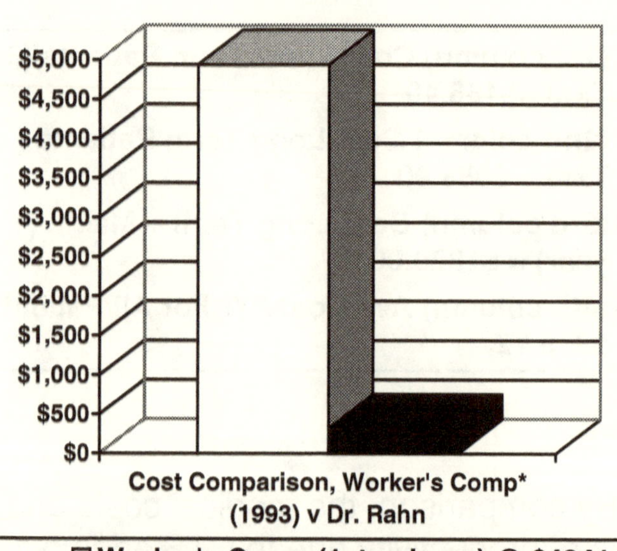

□ Worker's Comp (1st column) @ $4941.00
■ Dr. Rahn (2nd column) @ $318.12

* Workers' compensation statistics for the federal workforce. These statistics came from the U.S. Department of Labor's Office of Workers' Compensation

Programs as reported from October 1, 1993, through September 30, 1994 (8). This reflects a total of 185,927 claims with diagnoses of upper extremity disorder (7).

The average number of workdays lost for CTS in these statistics was 84 compared to 11.6 workdays lost by Dr. Rahn's patients on average per patient per year. Three patients were rated at total temporary disability and were out of work for a full year (patients '49', '93', and '100'). The first two patients underwent vocational rehabilitation and returned to full time employment. However, patient '100' is still on total disability. Patient '88' missed seven days and patient '89' missed 14 days.

Appendix A: Ergonomic Considerations

If your treatment is absolutely perfect and your symptoms stop completely, but your ergonomics are incorrect, your medical problem will return. The term ergonomics refers to the posture you use when you work. Typically there are two categories of ergonomics: Office ergonomics applies to everyone sitting at a desk, which often includes working with a computer keyboard and telephone. Industrial ergonomics applies to all types of physical labor, especially static positions such as assembly-line work. The main considerations in ergonomics are reach and neutral position.

First, let us discuss reach. The farther a person reaches, the greater the load on the body. If you pick up a large book, hold it close to your body, and then move it out to arms' length, you will quickly realize that the body does more work with the load at arms' length. The longer the reach, the more work you have to do. If you think of your arm as a lever and the book as a fulcrum, the longer the reach the more work is being done against you. If you still do not believe it, pick up a chair and hold it out at arms' length for a few minutes. After a minute or two your arms will shake with fatigue. The solution to the problem is to get as close as you can to the work being done.

At the same time there's an optimal height for a work surface so that you do not have to lean forward. Leaning forward actually creates the same lever effect on the body and uses a great deal of effort simply to support the unbalanced weight. Another consideration is how far you reach to move an object. To reduce your chance of injury, keep your body close to the object as you pick it up so as to avoid reaching out with your arms, keep the object close as you move it, and turn your toes in the direction you want to go. Remember, "The nose follows the toes". Along the same lines, the movements that create the greatest stress on a disk in the spine are bending, reaching, and twisting.

In order to understand the second major consideration in ergonomics, neutral position, think about how you might sit at a computer keyboard. The body is 'apparently' not doing any work at all. However, as mentioned earlier in this book, if a person elevates their shoulder a fraction of an inch, they are unknowingly doing the work of lifting the arm's weight. When a person has tension, the tendency is to raise or elevate the shoulders and arm. While doing workplace ergonomic evaluations in St. Croix, Dr. Rahn removed the arms of recently purchased $800 chairs to the dismay of the corporate executives. Though the employees felt they were resting their arms on the chairs, they were actually recruiting the muscles of the top of the shoulder to lift the weight. If this posture is held chronically, muscle spasm will lead to inflammation, which even in very small amounts can create fibrous tissue. The body wants to 'rope up' the weight of the arm with fibrous tissue to reduce the amount of energy and work required to hold up the arm. The net result is that the muscle tissue is less elastic, has decreased circulation, and decreased strength. This chronic tightness usually puts pressure on nerves and blood vessels going into the arm. Again, this diagnosis is thoracic outlet syndrome (TOS), which is the most common cause of carpal tunnel syndrome (CTS).

In summary, industrial workers should stay as near to their work as possible to decrease reaching and bending. As much as possible, workers should look straight ahead at the work without turning their heads for extended periods of time. In the office setting, the tool bar on the monitor should be level with the eyes, the monitor directly in front of the worker, and the focal distance to the monitor the same as the distance that person would hold a piece of paper for reading purposes. Monitors are very commonly set farther back than the optimal focal distance, which causes individuals to move their heads forward and therefore tighten up the muscles in the front of the neck and put pressure on the nerves going into the arm. Office chairs should be adjusted to create a 90° angle at the hips and knees so that the feet rest flat on the floor. If the keyboard cannot be lowered, a footrest is needed to maintain the 90-degree hip and knee angles. Most importantly, the shoulders need to be down all the way and elbows close to the sides of the body. The elbows should form a 90° angle. The hands should neither tip back nor down in relation to the forearms and should be in a neutral position. A wrist rest may be needed to accomplish this neutral wrist position. An office worker should not have to hold any material at all! Also, the mouse should be situated directly beside the keyboard to minimize any reaching.

Another consideration is eliminating chronic irritation by using a rotation schedule or switching between tasks. This is more easily accomplished in an industrial setting. Many times workers will stay at one particular task station for two hours and then move to a different job throughout the remainder of the day. However, even an office worker may be able to break up work by alternating between the computer, phone, paperwork, and filing.

An experienced healthcare provider, especially a massage therapist, can identify the ergonomic risk factor making your muscles tight and fibrotic. Of course, this also correlates with weak muscles and compression points on nerves. Dr. Rahn has traveled across the United States correlating the ergonomic cause to the actual physical presentation and treatment. This approach most certainly will yield very good results when addressing chronic workers' compensation problems. The cost of an on-site ergonomic evaluation is small in comparison to the cost of long-term treatment. Remember that the worker often cannot identify the muscle spasm creating the problem because the situation is chronic and the body has accommodated so that affected areas no longer hurt. Also, when nerve pain is involved, the area of pain, numbness or tingling is usually different than the site of nerve compression. If all ergonomic and treatment efforts fail to resolve the patient's symptoms, attention should be paid to the patient's stress level and diet, which are discussed in the next two appendix sections.

Appendix B: Stress Reduction Considerations

Even if someone's ergonomic set up is perfect and their diet is perfect, stress alone can create muscle spasm and all the problems associated with it. Adding to this is that usually the patients that have the greatest problems with stress are the ones that deny having a problem with stress. Perhaps subconsciously these people are afraid to address the issue, or they feel overwhelmed with an obscure topic. Like gravity, stress is a constant and so we seldom consider it specifically. Still, these people usually admit to difficulties with upset stomach, falling or staying asleep, tension headache, digestive ailments, and a myriad other problems.

Stress reduction can be approached in two modes. We can think of stress reduction as a simple exercise such as a breathing routine, visualization, or even a Hatha Yoga class. The other way to approach this, which is inherently important, is spiritually.

If I tell you not to think about a pink elephant, you will picture a pink elephant. Similarly, if I tell you not to think about paying the bills, you will think about paying the bills. Because of this tendency, we are unable to "not think about a problem." At the same time, we have an inability to think about nothing. Most people find it impossible to simply sit still and be quiet. Since we cannot think about nothing in order to relax, we must focus on something positive.

This topic may be more palatable for some people if we call this an exercise in affirmations. Those individuals who are comfortable with spirituality in general can adopt something from their spiritual process because a spiritual affirmation brings them a sense of reassurance.

Consider the Catholic rosary. There is the mechanical action of working the beads with the fingers that serves as a focal point. Then there is the repetition of a prayer, an affirmation, or something positive. We command the mind to be quiet and focus on the affirmation. If the mind wanders off to the old problem, we have to do the repetition twice as fast. Sometimes we even have to almost shout the mind down by speaking the repetition out loud.

If we practice this exercise 15 minutes a day for a period of months, we will see an increased ability to control our mind. Along with this we have less fear, anger and anxiety. When confronted with a problem or hardship we can quickly connect to our affirmation and calm ourselves. Eventually we are able to do the repetition slower and with more focus. After a degree of success with this practice, one may actually approach the practice of meditation. It is not the purpose of this book to teach meditation, so suffice it to say that 15 minutes a day spent in meditation or an affirmation exercise is effective in reducing stress and gaining the ability to minimize negative emotions.

Regular exercise and time spent in nature are very helpful as well. The overall effect of reducing stress will definitely help decrease muscle spasm, which brings with it decreased circulation, fibrous tissue, and ultimately compression of nerves and loss of function. Of course, diet can affect your stress level, which is the focus of the next section.

Appendix C: Dr. Rahn on Nutrition

My personal feelings about diet are that no supplement is as good as fresh, whole food prepared properly. I do recommend organic food. I also recommend a general multi-vitamin, multi-mineral supplement. Although many vitamins provide calcium in the form of calcium carbonate, I recommend other forms of calcium with a higher absorption, such as calcium lactate, calcium gluconate, and calcium citrate.

I believe that food loses nutritional value two hours after being cooked. Therefore, we should try to avoid 'manufactured' food. If you go to the store and buy a jar of pasta sauce and pasta, both items were previously cooked during preparation. Along the same lines, we should avoid preparing large amounts of food and eating leftovers days later.

To maintain an optimal level of blood sugar, the primary form of food should be complex carbohydrates. Because the germ in grain may be hard to digest, cracked grains or sprouted grains are best. Most markets sell a seven grain sprouted bread. Of course, bleached or white bread lacks the nutritional value of brown bread. Along the same lines, brown rice has more nutritional value than white.

Another common mistake is eating or drinking too much fruit or fruit juice with the thought that it is healthy. Fruit and fruit juice are packed with fruit sugars, which are very simple sugars. In large amounts, this can elevate blood sugar that then later falls below an optimal level. Because of this effect, fruit should be eaten in small amounts several times during the day. For example, one handful of grapes or half a large apple at one time is sufficient. These fruit sugars in moderation can help in stimulating the metabolic rate, which can help burn off fat.

Many common fad diets are based on eliminating carbohydrates and increasing protein. Proteins are hard to digest compared to carbohydrates. The metabolic end products of protein include sulfur and nitrogen compounds that, when elevated, can become toxic to the body. Some patients become so overloaded on protein that their perspiration actually has an ammonia smell to it.

My approach is to obtain a three-day diet diary including two weekdays and one weekend day. I look for what I identify as the greatest problem food and then try to eliminate or reduce this item with a healthy alternative that suits the patient's palate. This is the 'art' of nutritional counseling. Trying to change everything overnight usually fails. It is far better to change one item until that change becomes the norm and then go after another culprit. Food addictions are very strong, and changes are surprisingly difficult!

In short, the main component of the diet should be complex carbohydrates followed by moderate amounts of protein and several small servings of fruit or fruit juice throughout the day. While some fats and oils are essential for the body, these are present in whole foods, such as nuts, grains, eggs, and milk. Simply eliminating large amounts of salad dressing, fried foods, yellow cheese, and sweet desserts will lower our body fat.

An interesting reference for natural cures, by Kevin Trudeau, is cited (44).

As a health care provider, I would recommend for other providers to work with a company called Metagenics (35). I have used their detailed multi-page symptom survey to create a bar graph of relative stress on different organs. With training, this technique helps to make nutritional adjustments based on organ stress that is not acute enough to yield a positive lab test.

Another company I would highly recommend is Sunrider (36). I have used their products on and off for over 20 years. I utilize basic products such as the Sunbar and Vitashake. These are simply products created from whole food with the goal of providing the greatest amount of beneficial nutrients. If I eat one bar and have one shake per day my energy level and clarity of thinking increase. These effects are significant and patients consistently report having a similar benefit.

Sunrider is a multilevel marketing company. However, you may sign up to obtain product without attempting to create a business of signing up others and selling product. In the event a person would want to go on the Internet and sign up with me as their sponsor, my distributor number is **008 050 555**. My sponsor Margie Napolitano (37) is available to answer questions and network to check the availability of local meetings. If a patient wants to try a month's supply of the bar and shake before signing up with the company, we can also make arrangements to ship product. Any additional income generated from sponsoring others would help support our ongoing humanitarian efforts.

Appendix D: Technique for Neck Pain and Headache

The treatment protocol applied to carpal tunnel syndrome (CTS) is basically the same for neck pain and headache, although there may or may not be a specific nerve compression presentation.

Most people who complain of migraine headaches actually suffer from a muscle tension headache. They may like to use the term migraine because they feel a generic muscle tension headache could not possibly hurt so much. With a tension headache, there typically is compression of the greater and lesser occipital nerves due to muscle spasm of the base of the skull. Usually muscle tension patterns are accompanied by muscle spasm of the temporomandibular joint (TMJ) or jaw joint. Less frequently, muscle spasm and lymphedema compress the facial nerve where it crosses over the ramus of the mandible.

The same muscle spasm release described in the CTS or thoracic outlet syndrome is used where the outline of the spasm is palpated. The center of the spasm is then compressed, which is called 'ischemic compression'.

For the TMJ, this technique is applied to the masseter at the angle of the mandible, both pterygoid muscles, and the temporalis muscle. At times, Dr. Travel's trigger point 'spray and stretch' technique can be used over the ramus of the mandible to help reduce spasm (38). With this technique, the eye must be protected to prevent any contact between the skin refrigerant and the surface of the eye. Lymphatic drainage techniques are also utilized for the front of the neck. Upledger release is applied to the jaw (39). Lastly, an exercise is utilized in which the patient learns to relax muscle tension in the TMJ muscles by wiggling the jaw open and closed, by hand, with a broken rhythm. A broken rhythm is necessary because TMJ muscles are recruited if a regular rhythm is used.

Muscle spasm in the neck is usually the major contributing factor for either neck pain or headache. Various different specific massage therapy techniques and joint mobilization techniques are utilized for the neck. A gentle manual traction helps reduce muscle spasm, as does ultrasound and icing.

A commonly overlooked component of the headache presentation is dural tension. The dura is the thick membrane that wraps around the spinal cord and brain itself. Muscle spasm anywhere in the back can traction nerve roots, which transmits the tension up the spinal column to the base of the skull, center of the forehead or temples. Therefore, we have to consider three major contributing factors to headache: muscle tension, joint restriction, and dural tension. At times the dural presentation can create unexpected presentations including increased tone, muscle tension, on one half of the body. Another example of how this tension may be transmitted or transferred is when a person falls on the tailbone and the tailbone is actually tipped to the side, which results in the tension being transmitted all the way up to the point of pain were the dura attaches to the center of the forehead or temples.

The many specific applications of massage therapy technique and joint mobilization is too detailed and comprehensive a topic to be addressed here (40, 41, 42, 43, 44). However, the fundamental aspects of treatment, including basic deep therapeutic massage 'trigger point therapy' and non-thrust joint mobilization, almost always result in alleviation of the headache presentation without the use of medication.

Appendix E: Technique Applied to Leg Pain

The treatment protocol applied to carpal tunnel syndrome (CTS) is basically the same as the treatment for leg pain. With leg pain, the nerves typically compressed are the femoral nerve at the front of the hip in the inguinal canal (anterior femoral nerve entrapment syndrome) or the sciatic nerve at the back of the hip under the origin of the pyriformis muscle (pyriformis syndrome).

Most people complain that the leg is heavy, or that it is hard to walk. This results from the femoral nerve creating weakness of the front of the thigh (quadriceps muscle). Distal (toward the knee) atrophy is often noted. If the patient sits in a chair and places his or her hands on the top of the knee and then pulls the leg up, a weakness may be noted in one or both legs. For muscle tests, the patient should lie back (supine) with hips and knees both flexed 90 degrees. The physical therapist or chiropractor then provides resistance as the patient pulls one knee at a time toward his or her head.

A sciatic presentation may or may not have pain since chronic presentations may be without pain. There may be 'foot drop' (a weakness at the ankle) in which the patient has difficulty pulling the foot up. Changes in light touch may be noted (pinwheel) and there may be changes in deep tendon reflexes (DTRs). Even if there is a positive finding of discopathy (disc bulge) on radiographs, reducing the pyriformis spasm may greatly improve the degree of symptoms and objective findings.

The standard muscle spasm release is used where the outline of the spasm is palpated, and then the center of the spasm is compressed. This compression is called 'ischemic compression' and is performed on the pyriformis muscle for sciatic presentations and proximal hip flexors for femoral entrapment. The main finding at the front of the hip is typically lymphatic congestion. Palpation is usually tender. The hip and knee may be flexed with the foot on the treatment table to take the hip flexors off of 'stretch'. This deep soft tissue technique may be duplicated by the patient, provided they are cautioned that the femoral nerve should not be compressed against the bony edge of the inguinal canal.

For the sciatic compression, Dr. Travel's trigger point 'spray and stretch' technique (38) can be used over the pyriformis to help reduce spasm. Ultrasound is very effective, and we usually follow the ultrasound treatment with an ice pack. Lymphatic drainage techniques are utilized for the front of the hip. Also, release of the lower end (insertion) of the psoas is needed. With both sciatic and femoral decompression, pain and paresthesias usually result in the weeks following the first treatment. This is a result of nerve regeneration.

Stretches are provided for hip flexors and pyriformis according to Saunders (6). However, stretching is not initiated until the patient is out of the acute phase because traction on the nerve roots greatly irritates sciatic presentations. Usually, pelvic tilts are provided for low back stabilization along with low back stretching. Home care will include ice packs as needed for pain, except where contraindicated as with diabetes and certain nerve or circulatory problems.

Again, the many specific applications of massage therapy techniques and joint mobilization are too detailed and comprehensive to be addressed here (40, 41, 42, 43).

However, the fundamental aspects of treatment (including basic deep therapeutic massage 'trigger point therapy' and non-thrust joint mobilization) almost always result in alleviation of the leg pain presentation without the use of medication.

Appendix F: Dr. Rahn's Hand Exam Form

(DDX SH/ C/S)(1,2)

DDX Tests:

C/S Compression:

Cervical Percussion:

Adson's Test:

Shoulder Palpation (Muscle Spasm):

DTR:

Left:	**Bi**	**Tri**	**Forearm**
Right:	**Bi**	**Tri**	**Forearm**

Grip Strength Dynamometer: (station2 kg)

Settings: 1 2 3 4 5

Left:
Right:

Sustained Grip Strength
(station 2 / kg over seconds)

	start	15	30	45	60
Left:					
Right:					

Pinch Strength: (kg)

Digit	Major (3x)	Minor(3x)
II		
III		
IV		
V		

Lateral Strength:

Left (3times):

Right (3 times):

Decreased Hand ROM:

	(45-50)	(50-55)	(85-90)
Thumb Right: Flex:	**CMC**	**MP**	**IP**

	(0)	(0-5)
Ext : CMC	**MP**	**IP**

Abd (60-70)
Add (30)
Opp (to 5th)

(45-50)(50-55)(85-90)
Thumb Left: Flex: CMC MP IP

	(0)	(0-5)
Ext : CMC	**MP**	**IP**

Abd (60-70)
Add (30)
Opp (to 5th)

RIGHT (digits)	**(85-90)**	**(100-115)**	**(80-90)**
2nd	MP	IP	DIP
3rd	MP	IP	DIP
4th	MP	IP	DIP
5th	MP	IP	DIP

LEFT (digits)	**(85-90)**	**(100-115)**	**(80-90)**
2nd	MP	IP	DIP
3rd	MP	IP	DIP
4th	MP	IP	DIP
5th	MP	IP	DIP

Wrist:	**RIGHT**	**LEFT**
DF (70-90)		

PF (80-90)

UD (30-45)

RD (15)

Elbow:	**RIGHT**	**LEFT**

Ext
Flex

Circ:	**RIGHT**	**LEFT**

(Forearm, 2 inches prox wrist)

Forearm:	**RIGHT**	**LEFT**

Pronation (85/90)
Supination (85/90)

Hyper / Hypoesthesia: (digits) (LIGHT TOUCH)

Two Point Discrimination:

Thenar Atrophy:

Carpal Tunnel Height:

Flexion R L
Extension R L

Hand / Forearm Volumetric Measurement:

R

L

Orthopedic Tests:

Retinacular Ligt: (passive/PIP neutral/flex DIP/ if no ROM then + for capsule or collateral ligt // flex PIP then DIP – if ROM normal then capsule OK):

Linburg's Sign: (flex thumb and extend index / pain +)

Lunatotriquetral Ballottment Test: (increased lunate post-ant ROM+)

Murphy's Sign: (make a fist, 3rd metacarpal level with 2nd and 4th, + lunate dislocation)

Ligt Instability Test: (DIP valgus and varus compared to other hand)

Watson Test: (stabilize rad/ulna and post-ant ROM scaphoid for ligt)

Grind Test: (stab MCP and axial compression-rot + with pain DJD)

Tinels Sign (wrist): (tap over carpal tunnel + pain)

Phalen's (wrist flexion) Test: (reverse prayer – 1 minute / paresthesia from 1st to lat ring finger + median nn)

Egasa's Sign: (flex middle digit and move radially and ulnarly / unable + interossei = ulnar paralysis)

Shoulder ROM:

L flex	**ext**	**abd**	**add**
R flex	**ext**	**abd**	**add**
Int Rot: L	**R**		
Ext Rot:L	**R**		

Appendix G: Equipment

1-2" Micrometer for measuring carpal tunnel height. Mitutoyo America Corporation, 958 Corporate Blvd., Aurora, IL 60504 Phone: (630) 820-3334 & Fax: (630) 820-2530 (Purchased through 'Tyler Tool Company). **Important note:** The edges of the micrometer are sharp & must be filed off to prevent cutting the patient. Larger patients may require a 2-3" micrometer.

Baseline Hydraulic Pinch Gauge, 50 lb. and **Hydraulic Hand Dynamometer** 200lb. The Human Solution, 12417 River Bend Road, #12, Austin, TX 78732 Phone: (512) 432-8774

Wartenberg Pinwheel

Seiko 10 Memory **Stopwatch**

Gulick **Anthropometric Tape, a** weighted tape to create consistency in circumference measurements.

2000 ml Polypropylene **Graduated Cylinder & Volumetric Tank** for measuring volume displacement by United Plastics Corp., www.usplastic.com

Hand Exercisers: "Planet Waves Gripmaster" ball to squeeze and rubber band to extend. Find it at www.samedaymusic.com. Also, Hand Exercisers, rubber bands on squeeze device, from Life Solutions Plus Inc.

Appendix H: Carpal Tunnel Height Measuring Procedure

To measure the change in carpal tunnel height, place the wrist in flexion (refer to the illustration on page 54). This position will give you the greatest tunnel height. Place the 'top' of the 'C' on the micrometer in the groove in the palmer side of the base of the palm (over the flexor retinaculum) and adjust the other side of the micrometer to touch the 'highest' point on the dorsal side. The micrometer is set to 'click' when a standard pressure is achieved. Repeat this procedure with the wrist in extension.

If there is a significant difference, ligamentous laxity is indicated. This is in conjunction with the manual test mentioned above where the patient places their hand on a hard surface and the provider presses downward on the high point of the curve (convexity). A springy movement of 1/8 inch, or more, indicates ligamentous laxity. If the curve is 'collapsed' and this movement causes pain, the carpal(s) may be 'displaced or locked' toward the palmar side of the hand, and a joint manipulation is indicated.

Appendix I: Works Cited

(1) Cummins JT, Rahn CL, **Rahn Roger S**. Microscopic observations on endogenous fluorochromes within a nerve fiber excited by a 325 nm He-Cd laser, Journal of Microscopy 127 (Pt 3): 277-85, September, 1982. (Also presented at the National Neuroscience Convention, Atlanta, GA, 1980).

(2) Management of Common Musculoskeletal Disorders, Third Edition, Darlene Hertling and Randolph M. Kessler, Published by Lippencott-Raven 1996.

(3) Orthopedic Physical Assessment, Second Edition, Magee, W. B. Saunders Company, 1992, pp 179-213, examination and assessment of forearm and hand.

(4) Atlas of Human Anatomy, Frank H. Netter, Ciba Pharmaceuticals Division 1995.

(5) Stretching, Anderson, Shelter Publications, 2000, shoulder and hand pp 42-48, 88-89.

(6) Saunders Basic Exercises, ToolsRG, customizable patient exercise / stretching software, www.toolsrg.com.

(7) Sathya Sai Baba. Website: http://www.sathyasai.org/

(8) Lee CH, Kim TK, Yoon ES, Postoperative morphologic analysis of carpal tunnel syndrome using high resolution ultrasonography. Ann Plast Surg 54: 2, 143-6, Feb, 2005.

(9) Occupational Upper Extremity Disorders in the Federal Workforce: Prevalence, Health Care Expenditures, and Patterns of Work Disability. Journal of Occupational & Environmental Medicine. 40(6):546-555, June 1998. Feurstein, Michael PhD; Miller, Virgina L. MD; Burnell, Lolita M. PhD; Berger, Ruth BS.

(10) Carpal Tunnel Syndrome: Diagnosis and Treatment, Eur. Surg. 2003; 35:196-201. Hildegunde Piza-Katzer, Department of Plastic and Reconstructive Surgery and the Ludwig Boltzman Institute for Quality Control in Plastic and Reconstructive Surgery, University of Innsbruck, Austria.

(11) Facts about carpal tunnel syndrome, James W. Strickland, MD, American Academy of Orthopaedic Surgeons, http://www.aaos.org

(12) Carpal Tunnel Syndrome: A review of the literature with recommendations for further research, C.M. Gross, J.D. Loyd, A. Nelson, and R.A. Haslam

(13) Hand Evaluation: a New Test for Carpal Tunnel Syndrome, Annals of Plastic Surgery. 46(2):120-124, February 2001. Ahn, Duck-Sun MD, FRCS(C).

(14) MRI of the Carpal Tunnel: Implications for Carpal Tunnel Syndrome, Keir, Peter J.; Bower, Jason; Stanisz, Greg, Medicine and Science in Sports & Exercise: Volume 36(5) Supplement May 2004 p S287-S288.

(15) Median Nerve Compression Can Be Detected By Magnetic Resonance Imaging of the Carpal Tunnel, Horch, Raymond E. MD; Allman, Karl Heinz MD; Laubenberger, Jorg MD; Langer Mathias MD; Stark, G. Bjoen MD, Neurosurgery. 41(1):76-83, July 1997.

(16) Does carpal tunnel stenosis predict outcome in women with carpal tunnel syndrome? S. I. Bekkelund; C. Pierre-Jerome, Acta Neurologica Scandinavica, Volume 107 Issue 2 Page 102 – February 2003, doi:10.1034/j.1600-0404.2003.02093.x

(17) Relationship of carpal canal contents volume to carpal canal pressure in carpal tunnel syndrome patients. Yoshida A; Okutsu I. J. Hand Surg [Br]. 2004 June;29(3):277-80.

(18) Chiropractic manipulation in carpal tunnel syndrome. Valente R; Gibson H, Department of Chiropractic Principles and Practice, Cleveland Chiropractic College and Clinic, J Manipulative Physiol Ther 1994 May; 17(4):246-9.

(19) Comparative efficacy of conservative medical and chiropractic treatments for carpal tunnel syndrome: a randomized clinical trial. Davis Pt et al. J Manipulative Pysiol Ther. 1998 Jun;21(5):317-26.

(20) Outcome evaluation of the operative management of lumbar disc herniation causing sciatica. Fisher C; Noonan V; Bishop P; Fairholm D; Wing P; Dvorak M. J Neurosurg. 2004 Apr;100(4 Suppl Spine):317-24.

(21) Results of posterior cervical foraminotomy for treatment of cervical spondylitic radiculopathy. J P Grieve; N D Kitchen; A J Moore, H T Marsh. British Journal of Neurosurgery V 14 (1) Feb 1, 2000. pp 40-43

(22) Cauda Equina Syndrome Secondary to Lumbar Disc Herniation: A Meta-Analysis of Surgical Outcomes. Ahn, Uri MD; Ahn, Nicholas U. MD; Buchowski. Jacob M. MS; Garrett, Elizabeth S. PhD; Seiber, Ann N. RN; Kostuik, John P. MD. Spine, Volume 25(12) June 15, 2000, pp 1515-1522.

(23) The Natural History of Asymptomatic Thoracic Disc Herniations, Wood, Kirkham B. MD et al. Spine, 22(5):525-529, March 1, 1997.

(24) Cervical Disc Herniations, Robert Pashman, MD, eSpine, http://www.espine.com/diagnosis/_SI.html .

(25) Comparison of open carpal tunnel release and local steroid treatment outcomes in idiopathic carpal tunnel syndrome. Serpil Demirci et al. Rheumatology International, Volume 22, Number 1, May 2002, pp 33-37.

(26) A randomized clinical trial of oral steroids in the treatment of carpal tunnel syndrome: a long term follow up. M-H Chang et al. J of Neurology Neurosurgery and Psychiatry 2002;73:710-714.

(27) Long-term symptom outcomes of carpal tunnel syndrome and its treatment. DeStefano F et al. J Hand Surg [Am] 1997 March;22(2):200-10.

(28) Long-term results of carpal tunnel release. Nancollas MP et al. J Hand Surg [Br] 1995 Aug;20(4)470-4.

(29) Long-term analysis of patients having surgical treatment for carpal tunnel syndrome. Kulick MI et al. J Hand Surg [Am] 1986 Jan;11(1):59-66.

(30) Splinting vs surgery in the treatment of carpal tunnel syndrome: a randomized controlled trial. Gerritsen AA et al. JAMA. 2002 Sep 11;288(10):1245-51.

(31) Carpal Tunnel Syndrome. (pathophysiology). Nigel L Ashworth. eMedicine, http://www.emedicine.com/pmr/topic21.htm .

(32) How Serious is Carpal Tunnel Syndrome? Harvey Simmon, MD, Editor in Chief. (MIT). http://www.tifaq.org/articles/carpal_tunnel_syndrome-sep98-well-connected.html .

(33) Cumulative Trauma Disorders and Repetitive Strain Injuries: The Future. Melhorn,

J. Mark MD. Clinical Orthopaedics & Related Research. (351):107-126, June 1998.

(34) An evaluation of medical and chiropractic provider utilization and costs: treating injured workers in North Carolina. Journal of Manipulative and Physiological Therapeutics, Sept. 2004;27:442-8. And reviewed as, "Work Comp Study: Chiropractic Less Expensive, More Effective Than Medical Care" in Dynamic Chiropractic, Michael Devitt, November 18, 2004, Volume 22, Issue 24.

(35) Metagenics San Clemente, 100 Avenue La Pata, San Clemente, CA 92673, Phone: (800) 692-9400, Website: www.metagenics.com.

(36) Sunrider International, 1625 Abalone Avenue, Torrance, CA 90501, Phone: (310) 781-3808, Fax: (310) 222-9278, Website: www.sunrider.com.

(37) Dr. Rahn's Sunrider Sponsors: Margie and Don Napolitano, Clovis, CA, Phone: (559) 297-7840, E-mail: margiedoll@aol.com. Dr. Rahn is Distributor # **008 050 555.**

(38) Myofascial Pain and Dysfunction: The Trigger Point Manual, Vol 1 and 2. by Janet Travell.

(39) Craniosacral Therapy, John E. Upledger

(40) Finando, Trigger Point Therapy: Defined, pages 3-7, Steps pp 29 and 30.

(41) Riggs, Deep Tissue Massage: General Theories, pages 15-18, Stroke Strategies, pp 30-36, (also recommend technique pages 42-59).

(42) Chatiow, Trigger Point Therapy: Referred pain patterns, pages 78-84, Proprioceptive Adjustment, pp 131.

(43) Tappan, Lymphatic Drainage Massage (LDM): LDM, pages 268-276, Skin Rolling, pp 134 and 135.

(44) Kevin Trudeau, Natural Cures "They" Don't Want You To Know About. Alliance Publishing Group.

Appendix J: Stretches and Exercises for TOS and CTS

Personal Exercise Program

Research Protocol, TOS / CTS

Provided by : Roger Rahn

Date : 08/20/2005

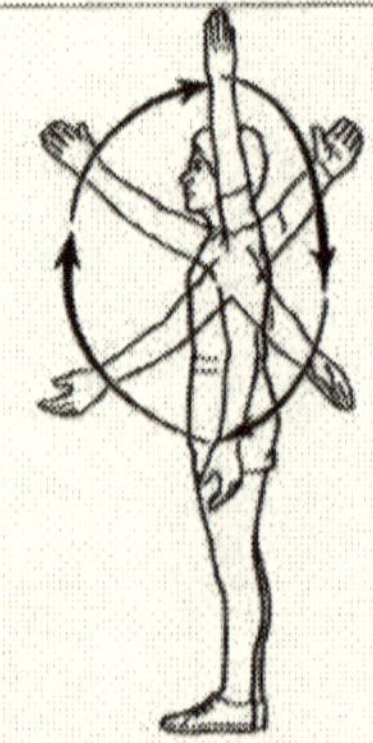

© The Saunders Group Inc.

1. Slowly make large circles backward with your arms
   6 circles thumb up & 6 circles thumb down
2. Now begin making forward circles
   6 circles thumb up & 6 circles thumb down
3. __2___ times per day

Built on Tools® RG 08/20/2005 1/4

1. Assume position shown, letting __each___ arm hang relaxed
2. Sway your whole body slowly to move arm forward and backward. Do not let the arm tense up - use only your body movement to begin the motion
3. Repeat, with the arm moving side to side
4. Repeat, with the arm moving in circular patterns, clockwise and counterclockwise
5 60 seconds, ___2__ times per day

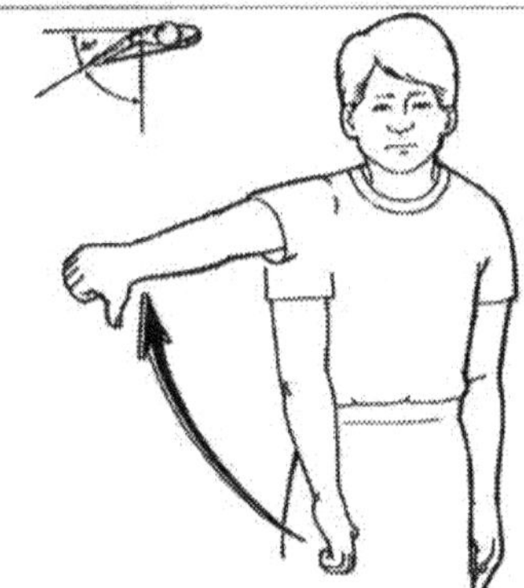

Lay on your back-
bend hips and knees, hook one knee over the other &
twist low back in one direction

1. Place opposite arm straight out with palm turned downward
2. Move arm higher or lower and feel for pull in front of shoulder
3. Hold __60___ seconds
4. ___1__ repetitions, __1__ times per day

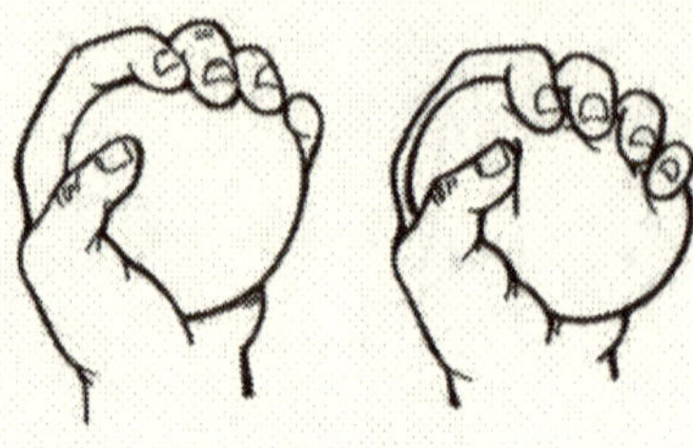

1. Hold a ball as shown with each hand
2. Squeeze as firmly as you can
3. Hold 3 seconds
4. 3 repetition, 3 times per day

caution: do not do any exercise that causes pain!

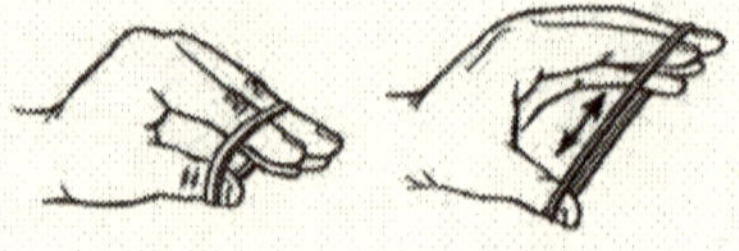

1. Place rubber band around fingers and thumb as shown
2. Move thumb and fingers apart as shown
3. Hold 3 seconds
4. 3 repetitions, 3 times per day

caution: do not do any exercise that causes pain!

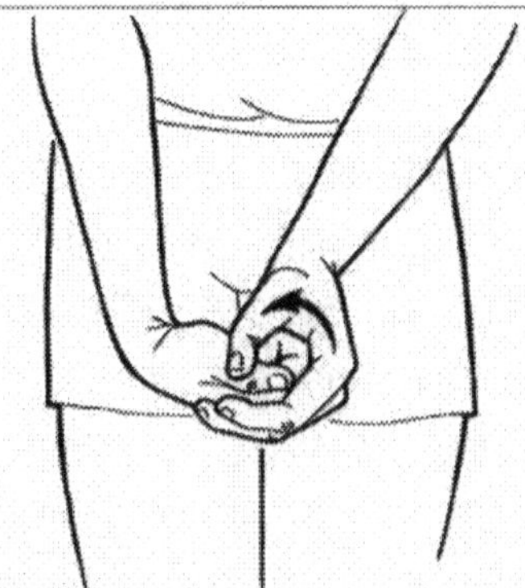

1. Hold __each__ wrist as shown, with the fingers closed
2. Bend the wrist until you feel a stretch
3. Hold __15___ seconds
4. ___1__ repetitions, ___3__ times per day

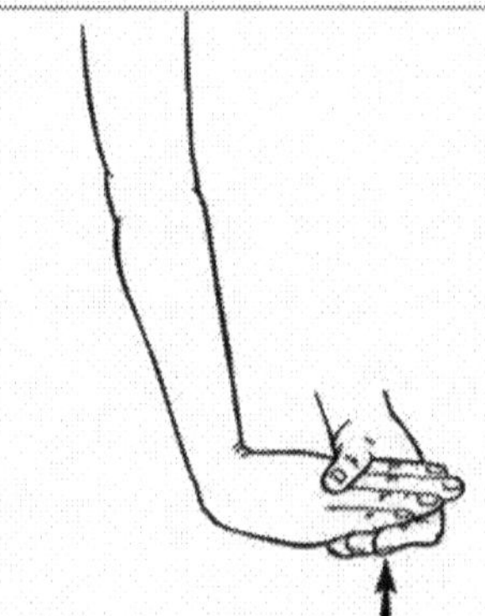

1. Hold __each__ wrist as shown, making sure to keep fingers straight
2. Bend the wrist and fingers upward until you feel a stretch
3. Hold __15___ seconds
4. __1__ repetitions, __3___ times per day

Appendix K: About Dr. Rahn

I knew I wanted in to be a doctor in the sixth grade. A teacher wanted us to write our epitaph. I wrote, "If in this lifetime I've made one person happier for one minute, then my life was successful."

When I began study at Los Angeles College of Chiropractic, I also received training from Santa Monica School of Massage. So, I became a massage therapist before I was a chiropractor. I graduated from Chiropractic College in 1988.

I can outwork anybody. The person I can thank for this work ethic is my father. From an early age I went out with my father to a construction site and I was free to lift as much weight and work as fast as I possibly could. Like most children I lived to please my parents so I helped with rough framing (nailing together the boards in the walls). As a child I could lift eight 2 x 4 studs. At age 11, I had a 28-ounce framing hammer and was allowed to hit the 16-penny nail two times, once to set it and once to sink it. We did piecework. This means that the more you did and the faster you worked, the more you were paid.

One time after starting college I needed some money. I drove to a construction site, evaluated what job on the site was in greatest demand and approached the foreman. Whereas it normally took two or three men two days to frame the walls of a tract house, I framed a house and stood the walls up by myself in eight hours. However, by my second year in college I had an ulcer. I knew how to work harder but I did not know how to work peacefully.

In 1991 I moved to a non-denominational meditation center in the Santa Monica Mountains and studied under the meditation master Swami Turiyasangitananda (also known as Alice Coltrane). We were taught to not judge others or comment on the process of others, as this is a form of judgment. Around the same time I made my first trip to India.

My teacher in America was the personification of devotion for God; my teacher in India was the personification of my work ethic, taught by my father. My Indian teacher, Sathya Sai Baba, is a true humanitarian with global works (7). I remember my father barking at me for turning around because it was wasted movement. I also remember my teacher in India admonishing us with the comment, "Hands that work are better than lips that pray." In short we had Christians, Buddhists, Muslims and others side-by-side, working together. I achieved synthesizing my hard work ethic with the peace of meditation. Meditation was not a state of consciousness to be left in the meditation seat; rather, this peace was to be carried with us throughout the workday.

There were many trips to India beginning in 1991 and my last trips spanned about two years, ending in 2002. I became a meditation teacher with students from all religions. I received criticism from family and friends because the humanitarian work that I felt called to do did not bring home a paycheck. My response was that if I was doing what I felt I was supposed to do and I was helping people, this should be 'okay'. Others responded that since I had friends who were not Christian, I must not be Christian. I think there were other factors at play, but suffice to say that a lot of my process was without support from friends and family.

At one time I had a Web site regarding meditation and spiritual practices. I feel that following the completion of this book I will write another book about my spiritual experiences. My motivation for both books is the same; I share whatever I can to help people. If I were to die tomorrow, I would know that I accomplished what I set out to accomplish in my sixth grade assignment, namely, I had the opportunity to help some people and decrease their suffering.

One other comment I'd like to make because it had such a profound influence on me is that in a conversation with a patient, the patient said, "Go and preach the gospel and, if you must, speak". To me this means 'do not judge, simply be an example.' I think this personifies my role. I am not here to convert and change others' viewpoints. I'm here to help. It sounds so simple, which it should be!

I hope this book has been of benefit. Of course, more research is needed as mentioned earlier. Your contributions, comments, references to literature, and suggestions are always appreciated.
Thank You!

The authors, Angela and Roger, with their son, Madhava, and daughter, Shraddha.

www.ingramcontent.com/pod-product-compliance
Lightning Source LLC
Chambersburg PA
CBHW022008170526
45157CB00003B/1198

* 9 7 8 1 4 1 1 6 4 4 3 1 1 *